ROCK 'N LEARN®

Addition & Subtraction ROCK

written and produced by:
Brad Caudle
Richard Caudle

musical performances by:
Brad Caudle

vocals by:
Brad Caudle
Jean July
Tom McCain

inside illustrations:
Bart Harlan

cover design:
Windy Polasek

Instructional Guide

Available in rock, rap, and country versions, Rock 'N Learn math programs have a cool sound that holds attention. These programs get results with students of all ages, from beginning learners to those with remedial needs.

At school, this program is perfect for group use or listening centers. At home, busy parents will love the way Rock 'N Learn programs allow independent practice. Less time needed for basic drill means more time to teach understanding and application of math concepts.

1. Before memorizing facts, learners need a clear understanding of what addition and subtraction accomplish. Effective teachers often introduce number concepts by using manipulative materials such as blocks, beans, or other objects. Point out practical applications of addition and subtraction, such as counting correct change.

2. Because this programs works like "musical flash cards," learners usually need the book in front of them. Rather than allowing students to write in the book, you may photocopy the pages for instructional use. Students solve the facts as they follow in the book and sing along.

3. Learning sessions need not be lengthy to be effective. Although the appropriate length depends on individual attention span, frequent sessions of about 15 minutes can produce good results. Students do best when they focus on one or two songs before advancing to other songs.

4. After memorizing the facts, learners should periodically listen to the entire recording to improve speed and maintain retention.

5. Unlike some math recordings, answers are never omitted by the performers. This is because students may think they know the answers when they actually do not. Always hearing the correct answer helps prevent incorrect associations. Also, research shows we learn best when we hear complete units of information. Encourage learners to say each entire fact, even if they must answer after the performer.

6. As an alternate activity, have students complete the work sheet for a song and then listen to the program to check answers.

7. In completing the math mazes, students may either solve each fact as they go along or solve all the facts and then work the maze.

Add A While

Numbers help us keep the beat
Not too fast, not too slow
The better you can add, my friend
The higher you will go

A number plus zero is always
    the same number
Just stay there
A number plus one is always
    the next number
Just count up

When you're adding to the music
You can do it easily
So come on now
And add a while with me

$$\begin{array}{r} 3 \\ +\ 0 \\ \hline 3 \end{array}$$

When they ask you for addition
They may say **plus** or **add** or **and**
But we know it's all the same thing
So just join in with the band

Yeah!

A number plus zero is always the same number
Just stay there
A number plus one is always the next number
Just count up

$$\begin{array}{r} 3 \\ +\ 1 \\ \hline 4 \end{array}$$

When you're adding to the music
You can do it easily
So come on now
And add a while with me

Practice adding zero and one. _____

$$\begin{array}{r} 2 \\ + \ 0 \\ \hline \end{array} \qquad \begin{array}{r} 8 \\ + \ 0 \\ \hline \end{array} \qquad \begin{array}{r} 3 \\ + \ 1 \\ \hline \end{array} \qquad \begin{array}{r} 2 \\ + \ 1 \\ \hline \end{array}$$

$$\begin{array}{r} 7 \\ + \ 1 \\ \hline \end{array} \qquad \begin{array}{r} 8 \\ + \ 0 \\ \hline \end{array} \qquad \begin{array}{r} 6 \\ + \ 0 \\ \hline \end{array} \qquad \begin{array}{r} 1 \\ + \ 1 \\ \hline \end{array}$$

$$\begin{array}{r} 3 \\ + \ 0 \\ \hline \end{array} \qquad \begin{array}{r} 9 \\ + \ 1 \\ \hline \end{array} \qquad \begin{array}{r} 0 \\ + \ 1 \\ \hline \end{array} \qquad \begin{array}{r} 4 \\ + \ 0 \\ \hline \end{array}$$

$$\begin{array}{r} 12 \\ + \ 0 \\ \hline \end{array} \qquad \begin{array}{r} 13 \\ + \ 1 \\ \hline \end{array} \qquad \begin{array}{r} 19 \\ + \ 0 \\ \hline \end{array} \qquad \begin{array}{r} 14 \\ + \ 1 \\ \hline \end{array}$$

$$\begin{array}{r} 18 \\ + \ 0 \\ \hline \end{array} \qquad \begin{array}{r} 1 \\ + \ 7 \\ \hline \end{array} \qquad \begin{array}{r} 0 \\ + \ 8 \\ \hline \end{array} \qquad \begin{array}{r} 1 \\ + \ 15 \\ \hline \end{array}$$

Adding Up

This is a song to sing while you're adding
Sing it along with me
Add up to 10, the numbers are easy
Soon you are gonna see

$$\begin{array}{r} 2 \\ +\ 2 \\ \hline \end{array} \qquad \begin{array}{r} 2 \\ +\ 4 \\ \hline \end{array} \qquad \begin{array}{r} 2 \\ +\ 7 \\ \hline \end{array} \qquad \begin{array}{r} 2 \\ +\ 8 \\ \hline \end{array}$$

$$\begin{array}{r} 2 \\ +\ 6 \\ \hline \end{array} \qquad \begin{array}{r} 2 \\ +\ 3 \\ \hline \end{array} \qquad \begin{array}{r} 2 \\ +\ 5 \\ \hline \end{array} \qquad \begin{array}{r} 3 \\ +\ 6 \\ \hline \end{array}$$

$$\begin{array}{r} 3 \\ +\ 3 \\ \hline \end{array} \qquad \begin{array}{r} 3 \\ +\ 2 \\ \hline \end{array} \qquad \begin{array}{r} 3 \\ +\ 4 \\ \hline \end{array} \qquad \begin{array}{r} 3 \\ +\ 7 \\ \hline \end{array}$$

$$\begin{array}{r} 2 \\ +\ 2 \\ \hline \end{array} \qquad \begin{array}{r} 3 \\ +\ 5 \\ \hline \end{array} \qquad \begin{array}{r} 3 \\ +\ 3 \\ \hline \end{array} \qquad \begin{array}{r} 4 \\ +\ 6 \\ \hline \end{array}$$

$$
\begin{array}{r} 4 \\ + 4 \\ \hline \end{array}
\qquad
\begin{array}{r} 4 \\ + 2 \\ \hline \end{array}
\qquad
\begin{array}{r} 4 \\ + 5 \\ \hline \end{array}
\qquad
\begin{array}{r} 4 \\ + 3 \\ \hline \end{array}
$$

$$
\begin{array}{r} 5 \\ + 5 \\ \hline \end{array}
\qquad
\begin{array}{r} 4 \\ + 4 \\ \hline \end{array}
\qquad
\begin{array}{r} 5 \\ + 4 \\ \hline \end{array}
\qquad
\begin{array}{r} 5 \\ + 2 \\ \hline \end{array}
$$

$$
\begin{array}{r} 5 \\ + 3 \\ \hline \end{array}
\qquad
\begin{array}{r} 5 \\ + 5 \\ \hline \end{array}
\qquad
\begin{array}{r} 6 \\ + 3 \\ \hline \end{array}
\qquad
\begin{array}{r} 6 \\ + 2 \\ \hline \end{array}
$$

$$
\begin{array}{r} 6 \\ + 4 \\ \hline \end{array}
\qquad
\begin{array}{r} 7 \\ + 2 \\ \hline \end{array}
\qquad
\begin{array}{r} 8 \\ + 2 \\ \hline \end{array}
\qquad
\begin{array}{r} 7 \\ + 3 \\ \hline \end{array}
$$

Adding up, (adding up numbers) numbers up to ten

Adding up, (adding up numbers) numbers up to ten

I said we are

Adding up, (adding up numbers) numbers up to ten

4

Take Away

Take away
Take away some numbers
Subtract today
Take a number from another number

You may find it's easier
    if you sing it like a song
You can learn subtraction
Everybody sing along

$$\begin{array}{r} 3 \\ -\ 3 \\ \hline 0 \end{array}$$

If you take a number from itself
    zero is what you'll get
If you take zero from a number
    the number stays the same

I said now

$$\begin{array}{r} 3 \\ -\ 0 \\ \hline 3 \end{array}$$

If you take one from a number
    it's just like counting down
You will find this is the same
    anywhere around

Take away, take away some numbers
Oh, subtract today
Take a number from another number
Oh, take away, take away those numbers
Take away

$$\begin{array}{r} 3 \\ -\ 1 \\ \hline 2 \end{array}$$

Practice subtracting zero, one and a number from itself.

4	4	4	7
− 0	− 1	− 4	− 0

5	4	5	5
− 5	− 4	− 0	− 1

9	9	6	6
− 9	− 1	− 6	− 0

Match the words with their sign. Draw a line to connect each one with the sign that means the same.

minus

add   +

plus

take away

and   ─

subtract

Keep Subtracting

$$
\begin{array}{r} 3 \\ -\,2 \\ \hline \end{array}
\qquad
\begin{array}{r} 5 \\ -\,3 \\ \hline \end{array}
\qquad
\begin{array}{r} 4 \\ -\,3 \\ \hline \end{array}
\qquad
\begin{array}{r} 6 \\ -\,4 \\ \hline \end{array}
$$

$$
\begin{array}{r} 4 \\ -\,2 \\ \hline \end{array}
\qquad
\begin{array}{r} 5 \\ -\,4 \\ \hline \end{array}
\qquad
\begin{array}{r} 5 \\ -\,2 \\ \hline \end{array}
\qquad
\begin{array}{r} 6 \\ -\,2 \\ \hline \end{array}
$$

$$
\begin{array}{r} 6 \\ -\,3 \\ \hline \end{array}
\qquad
\begin{array}{r} 6 \\ -\,5 \\ \hline \end{array}
\qquad
\begin{array}{r} 7 \\ -\,3 \\ \hline \end{array}
\qquad
\begin{array}{r} 7 \\ -\,6 \\ \hline \end{array}
$$

$$
\begin{array}{r} 7 \\ -\,4 \\ \hline \end{array}
\qquad
\begin{array}{r} 7 \\ -\,2 \\ \hline \end{array}
\qquad
\begin{array}{r} 8 \\ -\,2 \\ \hline \end{array}
\qquad
\begin{array}{r} 8 \\ -\,4 \\ \hline \end{array}
$$

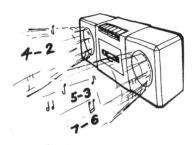

Keep subtracting
I know you can make it
We've got this chance to learn
And we're gonna take it

7 − 5	8 − 5	8 − 3	8 − 6
8 − 7	9 − 3	9 − 7	9 − 5
9 − 4	9 − 6	9 − 2	10 − 2
9 − 8	10 − 3	10 − 5	10 − 6
10 − 7	10 − 9	10 − 8	10 − 4

8   © 1993 Brad Caudle & Richard Caudle

Plus Minus Jam

We're gonna add some numbers, Man
Gonna subtract them, too
Numbers up to and from ten
Doin' them all for you

$$\begin{array}{r} 2 \\ + 7 \\ \hline \end{array} \qquad \begin{array}{r} 2 \\ + 3 \\ \hline \end{array} \qquad \begin{array}{r} 6 \\ - 3 \\ \hline \end{array} \qquad \begin{array}{r} 2 \\ + 2 \\ \hline \end{array}$$

$$\begin{array}{r} 2 \\ + 5 \\ \hline \end{array} \qquad \begin{array}{r} 5 \\ - 2 \\ \hline \end{array} \qquad \begin{array}{r} 6 \\ - 4 \\ \hline \end{array} \qquad \begin{array}{r} 3 \\ + 6 \\ \hline \end{array}$$

$$\begin{array}{r} 7 \\ - 5 \\ \hline \end{array} \qquad \begin{array}{r} 4 \\ + 4 \\ \hline \end{array} \qquad \begin{array}{r} 3 \\ + 3 \\ \hline \end{array} \qquad \begin{array}{r} 8 \\ - 4 \\ \hline \end{array}$$

$$\begin{array}{r} 4 \\ + 2 \\ \hline \end{array} \qquad \begin{array}{r} 8 \\ - 3 \\ \hline \end{array} \qquad \begin{array}{r} 9 \\ - 6 \\ \hline \end{array} \qquad \begin{array}{r} 9 \\ - 2 \\ \hline \end{array}$$

You're getting very quick, Man
In everything you do
See if you can guess the answers
Before I sing them to you

7	8	9	5
− 3	− 6	− 5	+ 3

4	9	5	4
+ 5	− 7	+ 5	+ 3

8	10	10	7
+ 2	− 3	− 5	+ 2

10	6	10	6
− 6	+ 4	− 8	+ 2

Well you did alright
Well you're doin' alright
Alright Man!

10

Double Time / Half Time ———————

Well, it's double time
Add a number to itself to make a double
Just sing along
Sing along with me and you will have no trouble

$$\begin{array}{r} 2 \\ + 2 \\ \hline \end{array} \qquad \begin{array}{r} 3 \\ + 3 \\ \hline \end{array} \qquad \begin{array}{r} 4 \\ + 4 \\ \hline \end{array} \qquad \begin{array}{r} 5 \\ + 5 \\ \hline \end{array}$$

$$\begin{array}{r} 6 \\ + 6 \\ \hline \end{array} \qquad \begin{array}{r} 7 \\ + 7 \\ \hline \end{array} \qquad \begin{array}{r} 8 \\ + 8 \\ \hline \end{array} \qquad \begin{array}{r} 9 \\ + 9 \\ \hline \end{array}$$

Now, see if you can sing the double before me...

$$\begin{array}{r} 6 \\ + 6 \\ \hline \end{array} \qquad \begin{array}{r} 3 \\ + 3 \\ \hline \end{array} \qquad \begin{array}{r} 9 \\ + 9 \\ \hline \end{array} \qquad \begin{array}{r} 8 \\ + 8 \\ \hline \end{array}$$

$$\begin{array}{r} 4 \\ + 4 \\ \hline \end{array} \qquad \begin{array}{r} 7 \\ + 7 \\ \hline \end{array} \qquad \begin{array}{r} 5 \\ + 5 \\ \hline \end{array} \qquad \begin{array}{r} 2 \\ + 2 \\ \hline \end{array}$$

We'll it's half time now
So let me show you how
To slice a number with math

If the answer is the same
As what you take away
You know you have found the half

$$
\begin{array}{cccc}
4 & 6 & 8 & 10 \\
-\ 2 & -\ 3 & -\ 4 & -\ 5 \\
\end{array}
$$

$$
\begin{array}{cccc}
12 & 14 & 16 & 18 \\
-\ 6 & -\ 7 & -\ 8 & -\ 9 \\
\end{array}
$$

Now, see if you can sing the half before me...

$$
\begin{array}{cccc}
14 & 10 & 6 & 12 \\
-\ 7 & -\ 5 & -\ 3 & -\ 6 \\
\end{array}
$$

$$
\begin{array}{cccc}
18 & 8 & 4 & 16 \\
-\ 9 & -\ 4 & -\ 2 & -\ 8 \\
\end{array}
$$

Secret Code _____

Oh no! Brad has lost his guitar and he has a concert to play. If
he solves the code, he will have a clue where to look. Help him
by working the problems and using your answers to solve the
secret code. Please hurry! The concert is tonight!

$$
\begin{array}{cccc}
2 & 5 & 7 \\
+3 & -3 & -5 \\
\hline
\end{array}
\qquad
\begin{array}{ccc}
9 & 6 & 4 \\
-6 & -5 & +2 \\
\hline
\end{array}
$$

☐ ☐ ☐   ☐ ☐ ☐

$$
\begin{array}{cc}
8 & 3 \\
-4 & +4 \\
\hline
\end{array}
$$

☐ ☐

Please help me find my guitar.

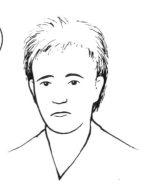

1	2	3	4	5	6	7
A	E	H	I	L	S	T

A-maze-ing Math

Help Brad get to Lee's house. When you come to a math problem, solve it. After writing your answer, keep drawing a line until the next problem, and so on.

If your answer is a **4** or a **7**, you are going the wrong way!

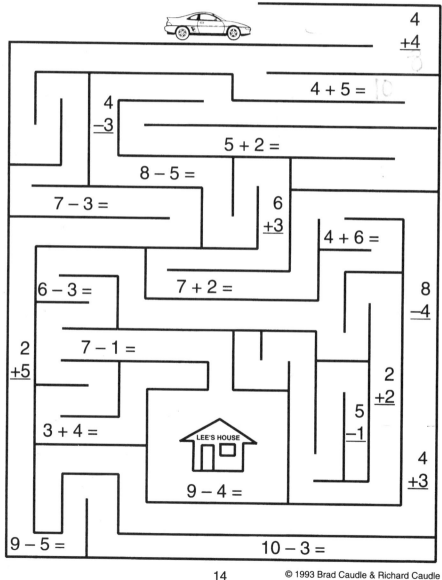

$$4$$
$$+4$$

$$4 + 5 = 10$$

$$4$$
$$-3$$

$$5 + 2 =$$

$$8 - 5 =$$

$$7 - 3 =$$

$$6$$
$$+3$$

$$4 + 6 =$$

$$6 - 3 =$$

$$7 + 2 =$$

$$8$$
$$-4$$

$$2$$
$$+5$$

$$7 - 1 =$$

$$2$$
$$+2$$

$$5$$
$$-1$$

$$3 + 4 =$$

LEE'S HOUSE

$$4$$
$$+3$$

$$9 - 4 =$$

$$9 - 5 =$$

$$10 - 3 =$$

Big Number Add _____

$$\begin{array}{r} 2 \\ + 9 \\ \hline \end{array} \qquad \begin{array}{r} 9 \\ + 7 \\ \hline \end{array} \qquad \begin{array}{r} 5 \\ + 9 \\ \hline \end{array} \qquad \begin{array}{r} 7 \\ + 8 \\ \hline \end{array}$$

$$\begin{array}{r} 8 \\ + 6 \\ \hline \end{array} \qquad \begin{array}{r} 6 \\ + 5 \\ \hline \end{array} \qquad \begin{array}{r} 7 \\ + 6 \\ \hline \end{array} \qquad \begin{array}{r} 8 \\ + 8 \\ \hline \end{array}$$

$$\begin{array}{r} 9 \\ + 3 \\ \hline \end{array} \qquad \begin{array}{r} 8 \\ + 6 \\ \hline \end{array} \qquad \begin{array}{r} 7 \\ + 8 \\ \hline \end{array} \qquad \begin{array}{r} 8 \\ + 4 \\ \hline \end{array}$$

$$\begin{array}{r} 7 \\ + 7 \\ \hline \end{array} \qquad \begin{array}{r} 5 \\ + 8 \\ \hline \end{array} \qquad \begin{array}{r} 7 \\ + 9 \\ \hline \end{array} \qquad \begin{array}{r} 4 \\ + 9 \\ \hline \end{array}$$

Add with me, the numbers are big
But you can get the answer right
Add quickly in your head
Get your math facts down tight

15

6 + 8	4 + 7	8 + 5	5 + 7
3 + 9	9 + 8	5 + 6	6 + 7
6 + 9	9 + 2	9 + 5	8 + 7
8 + 9	8 + 3	8 + 7	6 + 6
7 + 5	9 + 9	9 + 4	3 + 8
9 + 6	7 + 4	9 + 7	4 + 8

Having Fun _____

12	15	11	13
− 4	− 9	− 9	− 7

12	13	12	14
− 9	− 6	− 5	− 6

13	16	14	12
− 8	− 9	− 5	− 8

Havin' fun, havin' fun
You know that we're havin' so much fun

14	11	16	15
− 7	− 5	− 8	− 8

11	18	12	11
− 8	− 9	− 7	− 2

$$
\begin{array}{r}
17 \\
-\ 9 \\
\hline
\end{array}
\qquad
\begin{array}{r}
14 \\
-\ 8 \\
\hline
\end{array}
\qquad
\begin{array}{r}
11 \\
-\ 7 \\
\hline
\end{array}
\qquad
\begin{array}{r}
17 \\
-\ 8 \\
\hline
\end{array}
$$

We're havin' so much fun
Singing and subtracting
Music makes math easier
And it's not distracting

Havin' fun, havin' fun
You know that we're havin' so much fun

$$
\begin{array}{r}
12 \\
-\ 3 \\
\hline
\end{array}
\qquad
\begin{array}{r}
15 \\
-\ 8 \\
\hline
\end{array}
\qquad
\begin{array}{r}
13 \\
-\ 4 \\
\hline
\end{array}
\qquad
\begin{array}{r}
11 \\
-\ 4 \\
\hline
\end{array}
$$

$$
\begin{array}{r}
15 \\
-\ 7 \\
\hline
\end{array}
\qquad
\begin{array}{r}
12 \\
-\ 6 \\
\hline
\end{array}
\qquad
\begin{array}{r}
11 \\
-\ 3 \\
\hline
\end{array}
\qquad
\begin{array}{r}
14 \\
-\ 9 \\
\hline
\end{array}
$$

$$
\begin{array}{r}
16 \\
-\ 7 \\
\hline
\end{array}
\qquad
\begin{array}{r}
13 \\
-\ 5 \\
\hline
\end{array}
\qquad
\begin{array}{r}
11 \\
-\ 6 \\
\hline
\end{array}
\qquad
\begin{array}{r}
15 \\
-\ 6 \\
\hline
\end{array}
$$

Havin' fun, havin' fun
You know that we're havin' so much fun

These are Sum Colors!

Solve the equations. Color each part. If the sum is ...

11 or **14**, use **yellow**   **17** or **18**, use **blue**
12 or **13**, use **red**   **15** or **16**, use **green**

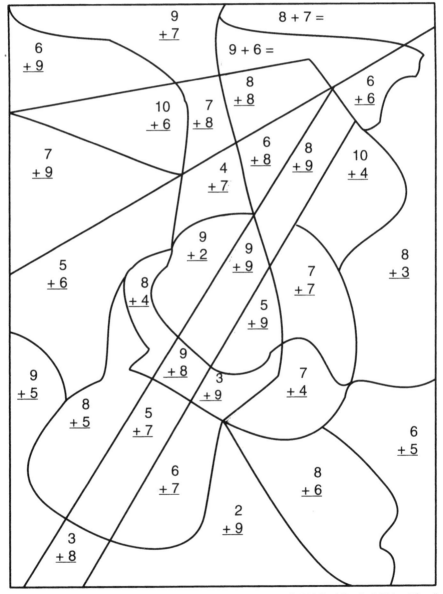

19

See the Difference! _____

Solve the equations. Color each part. If the difference is ...

5 or **6**, use **yellow**   **9** or **4**, use **blue**
7 or **8**, use **red**

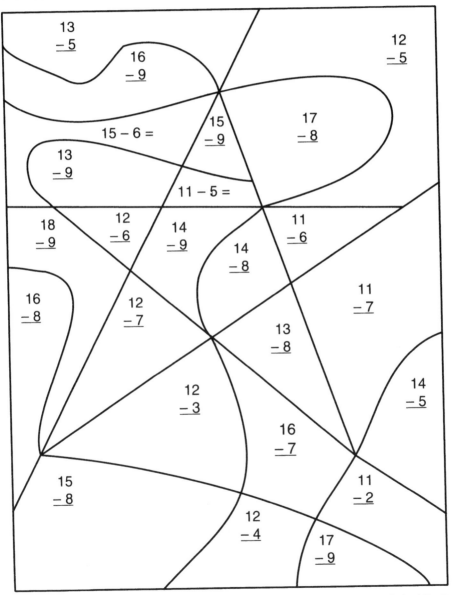

Everyone Rock _____

13 − 4	7 + 7	16 − 9	4 + 8
6 + 8	11 − 7	7 + 8	13 − 8
9 + 4	5 + 7	15 − 6	11 − 8
11 − 2	6 + 7	6 + 6	12 − 7

21

Everyone add, everyone subtract
Everyone rock 'n learn!
We're gonna sing to know our math
Many good grades we'll earn

$$
\begin{array}{r} 7 \\ + 9 \\ \hline \end{array}
\qquad
\begin{array}{r} 8 \\ + 9 \\ \hline \end{array}
\qquad
\begin{array}{r} 11 \\ - 3 \\ \hline \end{array}
\qquad
\begin{array}{r} 14 \\ - 8 \\ \hline \end{array}
$$

$$
\begin{array}{r} 8 \\ + 8 \\ \hline \end{array}
\qquad
\begin{array}{r} 11 \\ - 6 \\ \hline \end{array}
\qquad
\begin{array}{r} 14 \\ - 5 \\ \hline \end{array}
\qquad
\begin{array}{r} 5 \\ + 6 \\ \hline \end{array}
$$

$$
\begin{array}{r} 18 \\ - 9 \\ \hline \end{array}
\qquad
\begin{array}{r} 9 \\ + 5 \\ \hline \end{array}
\qquad
\begin{array}{r} 17 \\ - 9 \\ \hline \end{array}
\qquad
\begin{array}{r} 8 \\ + 7 \\ \hline \end{array}
$$

$$
\begin{array}{r} 13 \\ - 7 \\ \hline \end{array}
\qquad
\begin{array}{r} 9 \\ + 3 \\ \hline \end{array}
\qquad
\begin{array}{r} 12 \\ - 8 \\ \hline \end{array}
\qquad
\begin{array}{r} 9 \\ + 6 \\ \hline \end{array}
$$

Everyone Rock 'N Learn®!

Classroom Star

Come along with me
Everyone dance and sing
You'll be a star in the classroom
When you let the music ring

17 − 8	8 + 6	7 + 5	13 − 6
7 + 6	14 − 6	6 + 9	12 − 6
15 − 7	7 + 4	15 − 8	13 − 5
14 − 9	6 + 5	9 + 8	12 − 5

Oooh yeah!
Clap your hands to the numbers and the beat
Sometimes add, sometimes subtract
Oooh, it makes me move my feet!

9 + 2	3 + 9	13 − 9	8 + 3
15 − 9	12 − 3	8 + 5	12 − 9
9 + 9	2 + 9	16 − 7	16 − 9
4 + 9	16 − 8	11 − 4	5 + 9
17 − 8	8 + 4	14 − 7	12 − 4
9 + 7	15 − 7	12 − 3	3 + 8

Dot-to-dot Rock ────────────

Solve the equations, then connect the
dots in the order of your answers.

$8 - 8 =$ _____

$13 - 5 =$ _____ | $13 - 8 =$ _____ | $7 + 9 =$ _____

$8 + 6 =$ _____ | $9 + 9 =$ _____ | $12 - 2 =$ _____

$12 - 9 =$ _____ | $11 - 5 =$ _____ | $11 - 9 =$ _____

$18 - 6 =$ _____ | $8 + 9 =$ _____ | $15 - 8 =$ _____

$9 - 8 =$ _____ | $12 - 8 =$ _____ | $9 + 2 =$ _____

$8 + 7 =$ _____ | $7 + 6 =$ _____ | $16 - 7 =$ _____

25

© 1993 Brad Caudle & Richard Caudle

Concert Maze_____

Help Brad get to the concert. When you come to a math problem, solve it. After writing your answer, keep drawing a line until the next problem, and so on. If your answer is a **9** or a **15**, you are going the wrong way!

Start

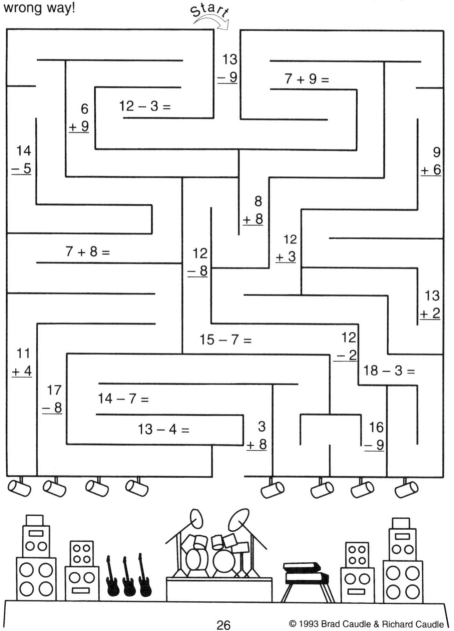

13
− 9

7 + 9 =

6
+ 9

12 − 3 =

14
− 5

9
+ 6

8
+ 8

7 + 8 =

12
− 8

12
+ 3

13
+ 2

15 − 7 =

12
− 2

11
+ 4

18 − 3 =

17
− 8

14 − 7 =

13 − 4 =

3
+ 8

16
− 9

26   © 1993 Brad Caudle & Richard Caudle

Answers

Page 2
2, 8, 4, 3
8, 8, 6, 2
3, 10, 1, 4
12, 14, 19, 15
18, 8, 8, 16

Page 3
4, 6, 9, 10
8, 5, 7, 9
6, 5, 7, 10
4, 8, 6, 10

Page 4
8, 6, 9, 7
10, 8, 9, 7
8, 10, 9, 8
10, 9, 10, 10

Page 6
4, 3, 0, 7
0, 0, 5, 4
0, 8, 0, 6

Page 7
1, 2, 1, 2
2, 1, 3, 4
3, 1, 4, 1
3, 5, 6, 4

Page 8
2, 3, 5, 2
1, 6, 2, 4
5, 3, 7, 8
1, 7, 5, 4
3, 1, 2, 6

Page 9
9, 5, 3, 4
7, 3, 2, 9
2, 8, 6, 4
6, 5, 3, 7

Page 10
4, 2, 4, 8
9, 2, 10, 7
10, 7, 5, 9
4, 10, 2, 8

Page 11
4, 6, 8, 10
12, 14, 16, 18
12, 6, 18, 16
8, 14, 10, 4

Page 12
2, 3, 4, 5
6, 7, 8, 9
7, 5, 3, 6
9, 4, 2, 8

Page 13
5 2 2   3 1 6
L E E   H A S

4 7
I T

Page 14

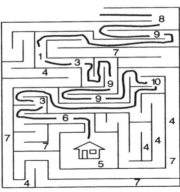

Page 15
11, 16, 14, 15
14, 11, 13, 16
12, 14, 15, 12
14, 13, 16, 13

Page 16
14, 11, 13, 12
12, 17, 11, 13
15, 11, 14, 15
17, 11, 15, 12
12, 18, 13, 11
15, 11, 16, 12

Page 17
8, 6, 2, 6
3, 7, 7, 8
5, 7, 9, 4
7, 6, 8, 7
3, 9, 5, 9

Page 18
8, 6, 4, 9
9, 7, 9, 7
8, 6, 8, 5
9, 8, 5, 9

Pages 19 & 20
see next page

Page 21
9, 14, 7, 12
14, 4, 15, 5
13, 12, 9, 3
9, 13, 12, 5

Page 22
16, 17, 8, 6
16, 5, 9, 11
9, 14, 8, 15
6, 12, 4, 15

27

© 1993 Brad Caudle & Richard Caudle

Page 19

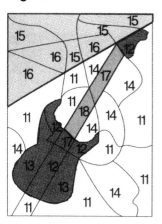

Page 20

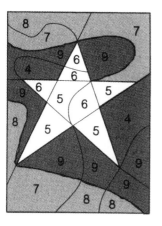

Page 21
9, 14, 7, 12
14, 4, 15, 5
13, 12, 9, 3
9, 13, 12, 5

Page 22
16, 17, 8, 6
16, 5, 9, 11
9, 14, 8, 15
6, 12, 4, 15

Page 23
9, 14, 12, 7
13, 8, 15, 6
8, 11, 7, 8
5, 11, 17, 7

Page 24
11, 12, 4, 11
6, 9, 13, 3
18, 11, 9, 7
13, 8, 7, 14
9, 12, 7, 8
16, 8, 9, 11

Page 25

		0
8	5	16
14	18	10
3	6	2
12	17	7
1	4	11
15	13	9

Page 26

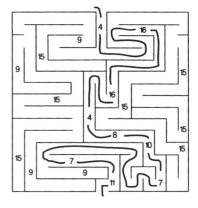

More Programs

Early Learning
Nursery Rhymes (CD/book, DVD or VHS)
Colors, Shapes & Counting (CD/book, DVD or VHS)
Alphabet (CD/book, DVD or VHS)
Alphabet Circus (DVD or VHS)
Alphabet Exercise (DVD or VHS)
Animals (DVD or VHS)
Getting Ready for Kindergarten (DVD or VHS)

Reading
Letter Sounds (CD/book, DVD or VHS)
Phonics (CD/book, DVD or VHS)
Phonics Easy Readers on DVD

Foreign Languages
Spanish I (CD/book), Spanish II (CD/book)
Spanish I & II (DVD or VHS)
French I (CD/book)

Math
Addition Rap (CD/book)
Subtraction Rap (CD/book)
Addition & Subtraction Rap (DVD or VHS)
Addition & Subtraction Rock (CD/book, DVD or VHS)
Addition & Subtraction Country (CD/book)
Multiplication Rap (CD/book, DVD or VHS)
Multiplication Rock (CD/book, DVD or VHS)
Multiplication Country (CD/book, DVD or VHS)
Division Rap (CD/book, DVD or VHS)
Division Rock (CD/book)
Money & Making Change (DVD or VHS)
Telling Time (DVD or VHS)
Beginning Fractions & Decimals (DVD or VHS)

& More!
Dinosaur Rap (CD/book)
Solar System (CD/book)
Presidents & U.S. Government (CD/book)
States & Capitals Rap (CD/book)
Grammar (CD/book)

Call 800-348-8445 or 936-539-2731
Or visit www.rocknlearn.com